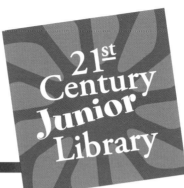

Honey Bee

by Katie Marsico

Published in the United States of America by Cherry Lake Publishing
Ann Arbor, Michigan
www.cherrylakepublishing.com

Content Adviser: The Entomological Foundation (www.entfdn.org)

Reading Adviser: Marla Conn, ReadAbility, Inc

Photo Credits: © Protasov AN/Shutterstock Images, cover; © Shaiith/Shutterstock Images, 4; © Lakeview Images/Shutterstock Images, 6; © dcwcreations/Shutterstock Images, 8; © Alex_187/Shutterstock Images, 10; © karunphol phanid/Shutterstock Images, 12; © angelshot/Shutterstock Images, 14; © Mirek Kijewski/Shutterstock Images, 16; © jadimages/Shutterstock Images, 18; © Quinn Martin/Shutterstock Images, 20

Copyright ©2016 by Cherry Lake Publishing
All rights reserved. No part of this book may be reproduced or utilized in any form or by any means without written permission from the publisher.

LIBRARY OF CONGRESS CATALOGING-IN-PUBLICATION DATA
Marsico, Katie, 1980-author.
  Honey Bee / by Katie Marsico.
     pages cm.—(21st century junior library) (Creepy crawly critters)
  Includes bibliographical references and index.
  ISBN 978-1-63362-591-4 (hardcover)—ISBN 978-1-63362-771-0 (pdf)—
ISBN 978-1-63362-681-2 (pbk.)—ISBN 978-1-63362-861-8 (ebook)
  1. Honey bee—Juvenile literature. I. Title. II. Series: 21st century junior library.
III. Series: Creepy crawly critters.

QL568.A6M525 2015
595.79'9—dc23
                                          2015005834

Cherry Lake Publishing would like to acknowledge the work of the Partnership for 21st Century Skills.
Please visit www.p21.org for more information.

Printed in the United States of America
Corporate Graphics

CONTENTS

5 Sweet Treat!

7 A Honey Bee's Body

13 What Happens in the Hive?

19 Humans and Honey Bees

22 Glossary

23 Find Out More

24 Index

24 About the Author

Honey bees produce the honey that humans eat.

Sweet Treat!

Have you ever tried honey on your toast? What made this sweet treat? There's a good chance honey bees were involved! These common bees are **insects** that produce honey.

Honey bees are covered in yellow and black fuzz.

A Honey Bee's Body

In many ways, honey bees look similar to other insects. They have three main body segments, or parts, and three pairs of legs. Honey bees also have two pairs of wings and a pair of **antennae** on their heads. Much of their body is coated in fuzzy hair.

In addition, honey bees have a stinger on the tip of their **abdomen**, or tail section. This stinger is used to stick **venom** into any

Bears like to eat honey, but they need to watch out for bees!

animals that the bees view as a threat. That includes people. Certain beetles, wasps, spiders, mites, toads, and birds are honey bee **predators**. So are mammals such as skunks, armadillos, anteaters, badgers, and bears.

How are honey bees different from other types of bees? And why do they die after they sting? These are good questions to ask an entomologist. An entomologist is a scientist who specializes in the study of insects.

Honey bees collect pollen from flowers.

Honey bees are not predators. Instead, their diet is made up of water and plant matter such as **pollen** and **nectar**. That's why you've probably spotted honey bees buzzing around your garden!

Look! Look at this photo of a honey bee. What colors do you notice? Are you able to spot all the features you just read about? What does it look like the bee in the picture is doing?

Honey bees live closely together in hives.

What Happens in the Hive?

Honey bees are **social** animals that live in nests called hives. The bees often build their hives in hollow trees. In some cases, as many as 80,000 bees exist together within a single hive! Their community features a female queen bee, female worker bees, and male bees, called drones.

The queen bee provides all of the eggs for the hive.

The queen bee **mates** with the drones and lays eggs in a honeycomb. This is a structure in the hive with six-sided cells, or tiny rooms. Eventually, baby honey bees, or larvae, hatch from the eggs.

The honeycomb serves another purpose, too. Worker bees use it to store honey.

Honey bees gather nectar and pollen from flowers, to feed the rest of the hive.

Honey-making begins with worker bees gathering nectar from flowers. Back at the hive, they use their mouths to pass the nectar to other workers. The workers then place the nectar in the honeycomb. It becomes honey as water **evaporates** from the nectar. Worker bees speed up evaporation by fanning their wings.

In a hive filled with as many as 80,000 honey bees, communication is important! These insects share information in several ways. Movement is one. Experts even describe certain honey bee motions as "dances"! Are you able to guess how else honey bees communicate?

Professional beekeepers need to wear safety gear.

Humans and Honey Bees

Honey bees feed on their honey, especially during winter months. It becomes an important source of energy when other food is hard to find. But people also enjoy this sweet treat. **Professional** beekeepers raise bees for their honey. People often use it as a spread, sweetener, and ingredient in various recipes.

Honey bees pollinate flowers, helping them to spread.

People and animals rely on honey bees to **pollinate** flowers, fruits, and vegetables. Unfortunately, honey bee populations are decreasing. There are several reasons why, including **pesticide** use. Luckily, many people are determined to figure out the best ways to save these incredible insects!

Honey bees do more than make honey and pollinate plants. They also produce wax to build honeycombs. People use beeswax to create wood polishes and candles. Are you able to think of any other products that trace back to honey bees?

GLOSSARY

abdomen (AB-duh-men) the rear section of an insect's body

antennae (an-TEH-nuh) sensitive organs on the head of some insects that are mainly used to feel, touch, and smell things

evaporates (i-VAP-uh-rates) changes into vapor or gas

insects (IN-sekts) small animals that have six legs, three body segments, and sometimes wings

mates (maytz) comes together with another animal to reproduce

nectar (NEK-tuhr) a sugary fluid produced by plants

pesticide (PESS-tuh-syde) a chemical used to protect plants or crops from pests, including insects

pollen (PAH-luhn) a fine, powdery substance made of tiny grains used in plant reproduction

pollinate (PAH-luh-nayte) to transfer pollen between plants, which often results in reproduction

predators (PREH-duh-turz) animals that kill other animals for food

professional (pruh-FESH-uh-nuhl) done as a paid job rather than a hobby

social (SOH-shuhl) preferring to spend time in a group, rather than alone

venom (VEH-nuhm) a naturally produced poison made by some animals

FIND OUT MORE

BOOKS

Huber, Raymond, and Brian Lovelock (illustrator). *Flight of the Honey Bee.* Somerville, MA: Candlewick Press, 2013.

Nelson, Robin. *From Flower to Honey.* Minneapolis: Lerner Publications, 2012.

Schuh, Mari. *Honeybees.* Minneapolis: Jump!, 2014.

WEB SITES

National Geographic Kids—Honeybee
http://kids.nationalgeographic.com/animals/honeybee/
Learn more fun facts about honey bees and other insects.

San Diego Zoo Kids—Bee
http://kids.sandiegozoo.org/animals/insects/bee
Review additional information and photographs related to honey bees.

INDEX

B
babies, 15
body, 7–11

C
communication, 17

D
drones, 13, 15

E
eggs, 14, 15

F
females, 13
food, 11, 16, 19

H
hives, 12, 13–17
honey, 4, 5, 8, 15, 17, 19
honeycomb, 15, 17, 21
humans, 4, 9, 19–21

M
males, 13
mating, 15

N
nectar, 11, 16, 17

P
pollen, 10, 11, 16

pollination, 20, 21
predators, 9, 11

Q
queen bees, 13, 14, 15

S
stinger, 7, 9

W
wax, 21
wings, 7, 17
worker bees, 13, 15, 17

ABOUT THE AUTHOR

Katie Marsico is the author of more than 200 children's books. She lives in a suburb of Chicago, Illinois, with her husband and children. Ms. Marsico would like to dedicate this book to Ms. Brown and her class at Edison School.